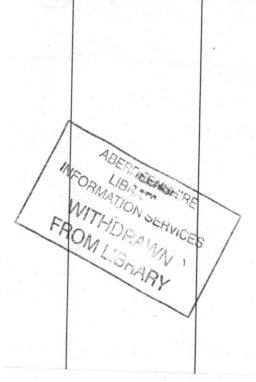

Are you a Grasshopper?

KINGFISHER
Kingfisher Publications Plc
New Penderel House
283–288 High Holborn
London WC1V 7HZ
www.kingfisherpub.com

First published by Kingfisher Publications Plc 2002

1 3 5 7 9 10 8 6 4 2

1TR/1201/TWP/GRS/150NYM

A CIP catalogue record for this book is available from
the British Library.

ISBN 0 7534 0552 0

Editor: Carron Brown
Series Designer: Jane Buckley

Printed in Singapore

Up the Garden Path

Are You a Grasshopper?

Judy Allen and Tudor Humphries

KING*f*ISHER

Are you a
grasshopper?

If you are, your mother
looks like this.

Your father looks
much the same.

Your mother laid her eggs
at the end of the summer.
She laid them deep in the grass,
just under the earth,
and covered them in frothy stuff.
The froth hardened into a pod
to keep the eggs safe.

You and your brothers
and sisters slept in
your eggs all winter –
but now it's spring.

Time to hatch.

All push together
to get out of the pod.

You look a bit like a
tiny worm, but
that's only because
you're wrapped in a
worm-shaped covering.

So unwrap yourself.

You're still tiny, and you haven't got
any wings, but now you look
like a grasshopper.

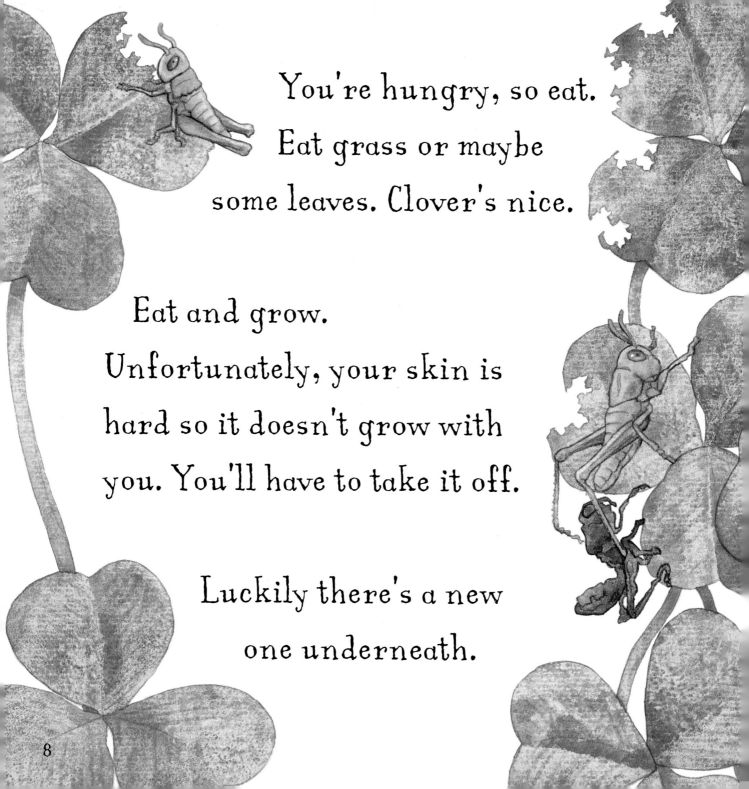

You're hungry, so eat.
Eat grass or maybe
some leaves. Clover's nice.

Eat and grow.
Unfortunately, your skin is
hard so it doesn't grow with
you. You'll have to take it off.

Luckily there's a new
one underneath.

Eat and grow and change your skin again. Now you have wing buds on your back.

Change your skin again.

Now you have small, stubby wings.

Eat and grow and change
your skin at least four times.

The final
skin-change
is the slowest and the biggest.

Struggle out of your old
skin and hang from

a grass stem

while your wings
grow to full size.

11

At last
you are a fully grown grasshopper.

You're bigger and stronger than before.
You have six legs, two pairs of wings,
two large eyes and two short feelers.

You have neat ear-holes on the sides
of your body, above your
back pair of legs.

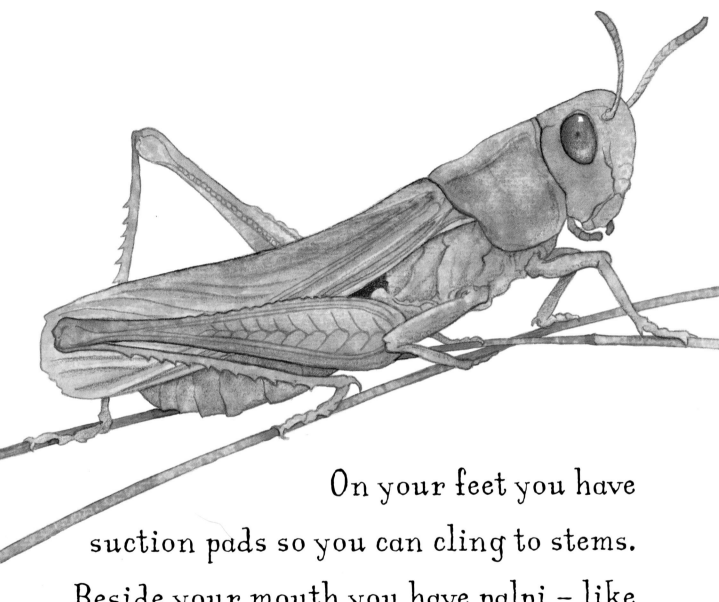

On your feet you have
suction pads so you can cling to stems.
Beside your mouth you have palpi – like
tiny fingers – to help push in your food.

This is a bush cricket.

You can tell it's a cricket
and not a grasshopper
because it has long feelers.

Don't stand around looking at it.
It may be eating a leaf right now,
but bush crickets eat
grasshoppers, too.

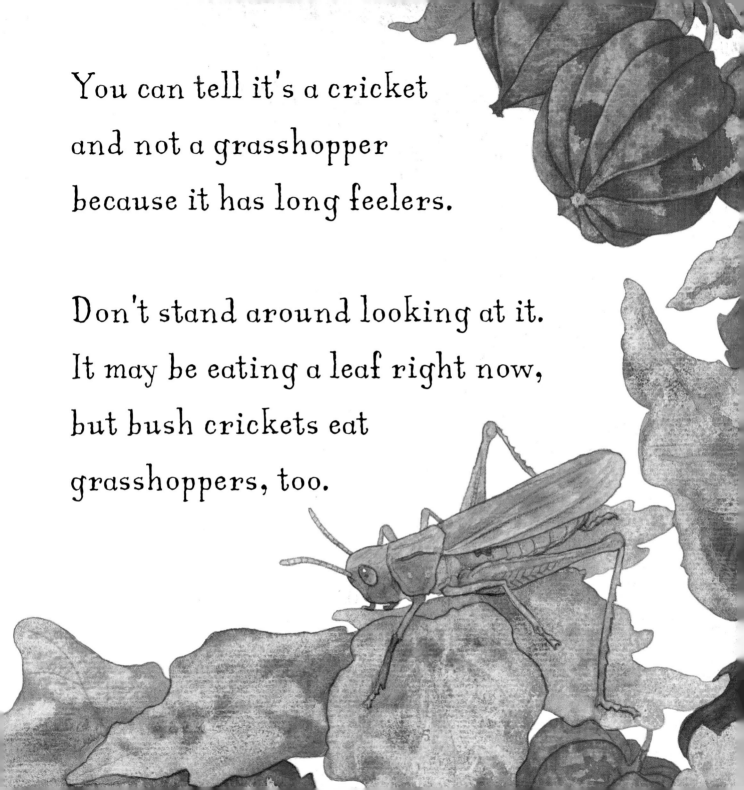

Don't get caught
in a spider's web.
Spiders eat grasshoppers.

Don't get too close
to a frog or a toad.
Frogs and toads
eat grasshoppers.

Watch out for birds.

Some birds eat
grasshoppers, too.

It's a dangerous
world out there
in the grass.

Always be ready
to escape from danger.

You could fly –
except you're
not very good at flying.

You could hide.

You're very difficult
to see if you creep
down among
the grass stems.

Or you could hop.
After all you ARE
a grasshopper.

19

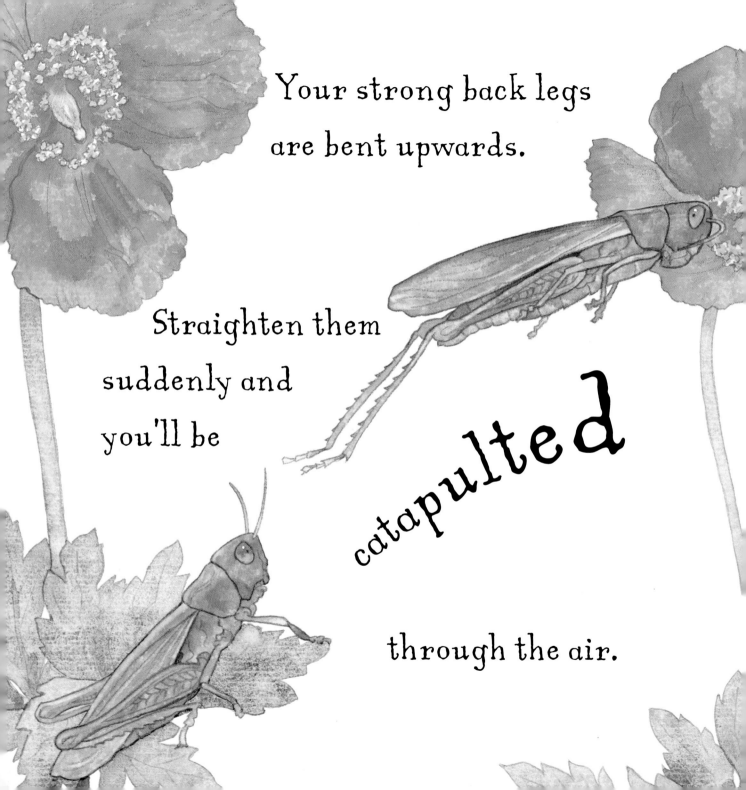

Your strong back legs
are bent upwards.

Straighten them
suddenly and
you'll be

catapulted

through the air.

You can jump amazingly
high and travel an
amazingly long way.

You're very light
and your skin is like a
suit of armour. You won't
hurt yourself when you land.

You have a row of tiny pegs on the inside of each leg. Pedal your legs up and down, fast. The pegs rub against your wings and make a ticking, chirping noise.

Keep going for about twenty seconds.

Have a rest.

Then start again.

If you are a male grasshopper, chirp loudly to attract a mate.

If you are a female grasshopper, you can chirp back if you want to.

The important thing is to mate and lay eggs for next year.

However, if your mother and father look like this

or this

or this

...you are not a grasshopper.

You are...

...a human child.

You haven't got wings.

You haven't got feelers.

You haven't got ears in
the sides of your body.

You probably
can't jump
amazingly high.

Never mind, you can do a great many things a grasshopper can't.

Best of all, you can make music
without pedalling your legs
up and down.

Did You Know...

...the grasshoppers in this book are
common green grasshoppers, but
there are more than 7,000 different
kinds of grasshopper. Also, they are
closely related to crickets, stick insects,
cockroaches and earwigs.

...a grasshopper can jump
20 times the length
of its own body.

...the locust is a large tropical grasshopper which is very good at flying. A swarm of locusts can eat a whole field of crops in one night.

...most grasshoppers die at the end of summer, after they've laid their eggs.